Detecting Malign or Subversive Information Efforts over Social Media

Scalable Analytics for Early Warning

WILLIAM MARCELLINO, KRYSTYNA MARCINEK,
STEPHANIE PEZARD, MIRIAM MATTHEWS

Prepared for U.S. European Command
Approved for public release; distribution unlimited

NATIONAL DEFENSE RESEARCH INSTITUTE

For more information on this publication, visit www.rand.org/t/RR4192

Library of Congress Cataloging-in-Publication Data is available for this publication.

ISBN: 978-1-9774-0379-7

Published by the RAND Corporation, Santa Monica, Calif.

© Copyright 2020 RAND Corporation

RAND® is a registered trademark.

Support RAND

Make a tax-deductible charitable contribution at
www.rand.org/giving/contribute

www.rand.org

Preface

This publication outlines the need for scalable analytic means to detect malign or subversive information efforts, lays out the methods and workflows used in detection of these efforts, and details a case study addressing such efforts during the 2018 World Cup. This information has operational relevance to the U.S. government, including the Department of Defense, and its allies tasked with agile detection and opposition of state-sponsored disinformation. It may also benefit researchers with an interest in going beyond post hoc human recognition of campaigns and social media companies interested in preventing their platforms from being co-opted by malign state actors.

This research was sponsored by U.S. European Command (USEUCOM) and conducted within the International Security and Defense Policy Center of the RAND National Defense Research Institute, a federally funded research and development center sponsored by the Office of the Secretary of Defense, the Joint Staff, the Unified Combatant Commands, the U.S. Navy, the U.S. Marine Corps, the defense agencies, and the defense Intelligence Community.

For more information on the RAND International Security and Defense Policy Center, see www.rand.org/nsrd/ndri/centers/isdp or contact the director (contact information is provided on the webpage).

Contents

Figures

Summary

The United States has a pressing capability gap in its ability to detect malign or subversive information campaigns in time to respond to them before these efforts substantially influence the attitudes and behaviors of large audiences. While there is ongoing research attempting to detect parts of such specific campaigns (e.g., compromised accounts, "fake news" stories), this report addresses a novel method to detect whole efforts.

As a proof of concept to detecting malign or subversive information campaigns over social media, we adapted an existing social media analysis method, combining network analysis and text analysis to map out, visualize, and understand the communities interacting on social media. This method works at scale. It allows analysts to look for patterns of disinformation in data sets that are too large for human qualitative analysis, reducing an enormous data set into smaller/denser data sets wherein a faint disinformation signal can be detected.

We examined whether Russia and its agents might have used the Russian hosting of the 2018 Fédération Internationale de Football Association (FIFA) World Cup as a launching point for malign and subversive information efforts. To do so, we analyzed approximately 69 million tweets—in English, French, and Russian—about the 2018 World Cup in the month before and the month after the Cup.[1] We were able to identify what appear to be two distinct Russian

---

[1] Specifically, we focused on tweets in a time frame of June 1–July 31, 2018.

information efforts, one aimed at Russian-speaking audiences, and one at French-speaking audiences:

- Russian-language efforts were characterized with relatively high confidence as likely including state-sponsored actions. These efforts were focused within two communities, and they centered on validating the Russian annexation of Crimea and justifying possible further military action against Ukraine.

- French-language efforts were also focused within two communities. One of these we rated with high confidence and the other with low confidence, as state sponsored. In one community these efforts centered on stoking right-wing extremism and populist resentment, and in the other community they centered on encouraging left- and right-wing extreme positions on immigration. [2]

Of particular note, the information effort in the French right-wing extremist community specifically targeted the anti–Emmanuel Macron, populist *gilets jaunes* (yellow vests) movement. Detecting this effort months before the *gilets jaunes* riots became English-language headlines illustrates the value of this method.

To help others use and further develop this method, we detail the specifics of our analysis in this report, including data, our workflows, and the specific analytic methods and algorithms used. We also share lessons learned during the course of the analysis, which stressed our existing technology infrastructure due to the large data set used. While the software, technology, and exact workflows are specific to the RAND Corporation, outside entities and researchers should be able to replicate the method in new contexts with new data sets.

---

2 We did not find indications of malign or subversive information campaigns in the English-language data.

Acknowledgments

The research team would like to thank COL Mike Jackson of ECJ39 for his support and encouragement in piloting this new approach to detecting malign information efforts. We would also like to thank a principal investigator of the study, Christopher Paul, for his support and confidence in an effort that involved significant risk of failure.

We would like to thank the original RAND research team that developed community lexical analysis: Elizabeth Bodine-Baron, Todd C. Helmus, Madeline Magnuson, and Zev Winkelman. We would also like to thank Michael Ryan and Joe Rybka for their technical support as we pushed RAND's software and technology infrastructure with by far the largest social media analysis we have conducted to date. Thanks also are due to Rouslan Karimov, who conducted the Twitter data pulls for the project. Lastly, we would like to thank Elizabeth Bodine-Baron and Olga Vartanova for their invaluable advice and insight on this document.

Abbreviations

CLA	community lexical analysis
FIFA	Fédération Internationale de Football Association
OM	Olympique de Marseille
TTPs	tactics, techniques, and procedures
WADA	World Anti-Doping Agency

Introduction: Detecting Malign or Subversive Information Efforts over Social Media

This publication explores the value of scalable analytic means to detect malign or subversive information efforts, lays out methods and workflows for detection of these efforts, and details a proof-of-concept study addressing malign information efforts detected around the 2018 Fédération Internationale de Football Association (FIFA) World Cup. Overall, our results suggest that organizations should consider adopting this or a similar social media analysis method to provide early warning to planners by detecting shaping actions over social media.

By *malign or subversive information efforts* we mean sustained efforts to spread false information or to manipulate information, with the political intention to affect perceptions and behavior.[1] Within this broader understanding of efforts, we may also refer to specific campaigns with specific objectives and

Pilot Outcome

We identified two likely Russian malign information campaigns piggybacking off the 2018 FIFA World Cup: one in Russian and one in French. Of particular note, the French-language information effort we detected involved stoking domestic populist tensions and the *gilets jaunes* movement. Early warning of the information effort was available months before these protests became headlines in English-language papers by November 2018.

---

[1] Miriam Matthews et al., *Russian Social Media Influence: Understanding and Defending Against Russia's Malign and Subversive Information Efforts in Europe*, Santa Monica, Calif.: RAND Corporation, RR-3160-EUCOM, forthcoming.

characteristics. Detecting these efforts and campaigns has operational relevance to the U.S. government, including the Department of Defense and its allies tasked with agile detection and opposition of state-sponsored disinformation. It may also benefit researchers with an interest in going beyond post hoc human recognition of malign campaigns to what we call in-time detection: short of real time, but rapid enough to respond. This method may also be of interest to social media companies interested in preventing their platforms from being co-opted by malign state actors.

Closing the Barn Door Too Late

The United States and its allies have an urgent need for better early warning of campaigns involving malign information (meant to harm) or subversive information (meant to undermine). Detecting and understanding these campaigns after they have run their course can provide valuable post hoc insight, but this delayed analysis does not permit the implementation of more rapid, proactive measures to address these efforts. Perhaps the best example of this to date was Russian activity around the U.S. 2016 presidential election. In hindsight, it was clear that Moscow had engaged in "a sweeping and sustained social influence operation consisting of various coordinated disinformation tactics aimed directly at U.S. citizens, designed to exert political influence and exacerbate social divisions in U.S. culture."[2] This information campaign was not, however, detected early enough for effective countermeasures.

This long-standing U.S. capability gap in detecting malign information efforts was a critical vulnerability during the War on Terror,[3] but has become even more urgent now in response to malign state actor

2   Renee DiResta et al., *The Tactics & Tropes of the Internet Research Agency*, New York: New Knowledge Organization, 2018, p. 4.

3   William D. Casebeer, "Narrative Technology to Detect and Defeat Adversary Ideological Influence," in Mariah Yager, ed. *SMA White Paper: What Do Others Think and How Do We Know What They Are Thinking?* Washington, D.C.: U.S. Department of Defense and Joint Chiefs of Staff, March 2018, p. 129.

efforts such as those of Russia.[4] Russia enjoys a significant advantage over the United States in this area. Because Soviet-era Russian information warfare science could not match that of the West technologically and monetarily, the Soviet, and later Russian, governments made deep investments in thinking about how to achieve informational affects. In contrast, U.S. work has been more focused on developing advanced technology and expanding infrastructure and capacity and has devoted relatively less attention to information theory.[5] Now that technological gaps have closed, the Russians' head start in informational theory and achieving informational effects makes them well positioned to conduct effective malign and subversive information efforts against the West.[6]

Detecting Malign Information Efforts as Wholes Rather Than Parts

Because detecting these efforts is such an urgent need, there is considerable work being done in the United States and the West to attempt to detect important elements of malign and subversive information efforts. Machine learning methods such as topic modeling–based approaches are being used to detect narrative themes and misinformation.[7] Other research focuses on dissemination methods and maneuvers. These include, for example, bot detection, bot network

---

[4] This is not to say that other state actors such as China and Iran are not also of concern.

[5] Timothy L. Thomas, "Dialectical Versus Empirical Thinking: Ten Key Elements of the Russian Understanding of Information Operations," *Journal of Slavic Military Studies*, Vol. 11, No. 1, 1998, pp. 40–62.

[6] We note that Russian informational activities, or *informatsionnoye protivoborstvo*, are not comparable to U.S. doctrinal concepts of information operations. *Informatsionnoye protivoborstvo* is a full integration of informational effects across military operations within a whole-of-government approach and has no real analog in U.S. doctrine or practice.

[7] P. A. Chew and J. G. Turnley, "Understanding Russian Information Operations Using Unsupervised Multilingual Topic Modeling," in Dongwon Lee, Yu-Ru Lin, Nathaniel Osgood, and Robert Thomson, eds., *Social, Cultural, and Behavioral Modeling: 10th International Conference, SBP-BRiMS 2017, Washington, DC, USA, July 5–8, 2017, Proceedings,* Cham, Switzerland: Springer International Publishing, 2017, pp. 102–107; Mikael Svärd and Philip Rumman, "Combating Disinformation: Detecting Fake News with Linguistic Models and Classification Algorithms," thesis, Stockholm: KTH Royal Institute of Technology, 2017.

structures,[8] and detection of hijacked (authentic) social media accounts as a distribution vector.[9]

However, all of this research is at the level of individual items, aimed at detecting essential parts of a larger whole: compromised accounts, "fake news" stories, specific themes, and the like. *Our research makes a contribution by moving to the level of aggregates to detect whole efforts.* That is, instead of looking at individual accounts we look at large groups of accounts that frequently interact (social media communities), and instead of individual tweets we look at very large collections of tweets grouped by community. By using network and text analytics, our analysts were able to visualize and understand social media discussion (and malign efforts to influence those arguments) at nonhuman scales.

Criteria for Russian Involvement

Our proof of concept for detecting malign and subversive information efforts specifically involved attempting to detect possible Russian efforts. Direct attribution for such efforts can be difficult, and so we developed criteria for Russian involvement. This included

1. alignment with Russia's documented informational objectives in Europe:[10]

   a. influencing public perception in other nations to support Russia's specific policy goals

8   Samer Al-Khateeb and Nitin Agarwal, "Understanding Strategic Information Manoeuvres in Network Media to Advance Cyber Operations: A Case Study Analysing Pro-Russian Separatists' Cyber Information Operations in Crimean Water Crisis," *Journal on Baltic Security*, Vol. 2, No. 1, 2016, pp. 6–27.

9   David Träng, Fredrik Johansson, and Magnus Rosell, "Evaluating Algorithms for Detection of Compromised Social Media User Accounts," in *ENIC '15: Proceedings of the 2015 Second European Network Intelligence Conference*, Washington, D.C.: IEEE Computer Society, 2015, pp. 75–82.

10   Matthews et al., forthcoming.

   b.   influencing public perception to further Russia's broader worldview and interests

   c.   exacerbating social tensions in other nations and spreading mistrust and confusion

2.   the presence of known Russian tactics, techniques, and procedures (TTPs) for informational efforts:

   a.   themes and messages supporting Russian policy ends and worldview, which can include broader themes such as general mistrust of the United States or the West, or more specific themes such as accusations that French president Emmanuel Macron is secretly gay or beholden to Saudi or Jewish interests

   b.   astroturfing/boosting known Kremlin-controlled social media accounts, or state-sponsored Russian media such as RT (formerly Russia Today) News and Sputnik

   c.   high use of fake social media accounts; these accounts could be bots (fully automated) or trolls (inauthentic personae accounts that are either human run or human run with machine assistance) but have in common deception about their authenticity.

Limitations on Attribution

Meeting some or all of the above criteria does not mean absolute certainty that an information effort is Russian in origin. We acknowledge that other actors might have aligned or overlapping interests, and that Russia does not uniquely own any of the TTPs listed above. We therefore accept a level of uncertainty, and instead focus on how closely informational efforts match the above criteria. At minimum, we felt that any set of activities within a given community must fulfill both of the top-level criteria above: aligning with Russian objectives and matching Russian TTPs. From there the number of TTP matches would increase the likelihood that we had discovered Russian malign activity. For example, in a case in which we discovered talk that furthered Russia's broader worldview (alignment) and included Russian media astroturfing (TTPs), we would have met a minimum threshold but have low confidence. But if we also found other TTPs (e.g., troll account indicators and specific known themes), then our confidence would be higher.

Case Study: The 2018 Fédération Internationale de Football Association World Cup

As an illustrative case study for our proposed method, RAND researchers conducted analyses of tweets that users disseminated via Twitter, the online social networking service. We focused on tweets that appeared to contain or reference content involving the 2018 FIFA World Cup, an international men's soccer tournament held once every four years. The 2018 World Cup was held in Russia during the time of this research, so it provided a somewhat unique opportunity to examine messaging disseminated during a large international event that clearly involved this country. As such, we could explore whether and how message content involving this event might have been used to promote Russian interests.

More broadly, relatively limited consideration has been given to how a country might use an international sporting event to frame the international conversation regarding a country's international status or political actions, such that many have previously assumed that "sport and anything in the political domain are distinct."[11] However, researchers are increasingly recognizing and examining the extent to which countries use large sporting events, such as the World Cup and the Olympic Games, as tools to communicate certain images regarding a country or to promote policy preferences and interests.[12]

[11] J. Simon Rofe, "Introduction: Establishing the Field of Play," in J. Simon Rofe, ed., *Sport and Diplomacy: Games Within Games*, Manchester, England: Manchester University Press, 2018, pp. 1–10.

[12] Suzanne Dowse, "Mega Sports Events as Political Tools: A Case Study of South Africa's Hosting of the 2010 FIFA World Cup," in J. Simon Rofe, ed., *Sport and Diplomacy: Games Within Games*, Manchester, England: Manchester University Press, 2018, pp. 70–86; Umberto Tulli, "'They Used Americana, All Painted and Polished, to Make the Enormous Impression They Did': Selling the Reagan Revolution Through the 1984 Olympic Games," in J. Simon Rofe, ed., *Sport and Diplomacy: Games Within Games*, Manchester, England: Manchester University Press, 2018, pp. 223–242.

Specific to Russia and its potential malign information campaigns, we considered whether prior Russian actions around major sporting events (i.e., the 2008 Olympics and the Russian invasion of Georgia and the 2014 Sochi Olympics and the invasion of Crimea) constitute a deliberate pattern.[13] If so, social media conversations around the 2018 World Cup would be a likely place to look for malign Russian efforts. Our analyses, therefore, also contribute to a growing area of research that considers sporting events as an element of international affairs, addressing a recent instance in which this might be observed.

The remainder of the report details our study and how to replicate it. Chapter Two lays out the data, methods, and workflows used in the study. Chapter Three consists of a detailed case study using scalable detection methods to look for Russian information efforts around a major sporting event. Chapter Four charts a way forward and gives recommendations. Finally, the Appendix details findings from the analysis that are not directly relevant to the study goal of detecting malign information efforts but may still be of interest to readers, as well as providing fuller context.

---

[13] For example, Russia may have timed its 2008 invasion of Georgia in part to take advantage of world attention on the Olympics; see Lionel Beehner et al., *Analyzing the Russian Way of War: Evidence from the 2008 Conflict with Georgia*, West Point, N.Y.: Modern War Institute at West Point, March 20, 2018. However, other scholars have argued that Russia is more likely to use sporting events to signal to internal rather than external audiences; see Jonathan Grix and Nina Kramareva, "The Sochi Winter Olympics and Russia's Unique Soft Power Strategy," *Sport in Society*, Vol. 20, No. 4, 2017, pp. 461–475.

CHAPTER TWO

Analytic Methods: A Template for Detecting Malign or Subversive Information Campaigns

Our goal in this report is to share our method so that others with an interest in supporting free, democratic discourse in digital spaces can use and further develop it. In this chapter we lay out our research plan, a top-level summary of our workflows, and more detailed descriptions of the data collection and analytic methods. While we used RAND's proprietary software in this analysis, in principle, anyone with access to network analysis algorithms, text analysis software, and network visualization software should be able to replicate this method. And while Twitter is well suited as a data source to this sort of effort, any interactive social media—any source where there is both text data for content and metadata indicating a connection and direction of connection—can be used.[1] Finally, this chapter includes technical details that may not be useful for general readers. However, for users hoping to replicate our method, we have included these details in the interests of completeness.

Overview: A Scalable Analytic Method for Early Detection

As a proof of concept, RAND researchers field-tested a new approach to detecting malign or subversive campaigns over social media, adapting an existing social media analysis method called community lexical

---

[1] Other examples would be Russia's VK service or China's Sina Weibo microblogging platform.

analysis (CLA), which combines network analysis and text analysis to map out, visualize, and understand the social communities interacting on social media. CLA works at scale and thus allows analysts to look for patterns associated with information efforts in data sets that are too large for human qualitative analysis.

CLA is a kind of data reduction, dividing an enormous data set into smaller/denser data sets wherein a faint signal might be detected. For example, imagine trying to detect hostile, state-sponsored operatives who plan to spread disinformation at a massive public event—monitoring thousands of voices all at once would be impossible. But if you could pull out the major conversations from the most influential people and analyze one conversation at a time, a faint signal that might be lost in the vast crowd may be detectable.

Bearing this analogy in mind, CLA works by

- using network analysis to resolve a large social media data set into lots of smaller "community" conversations
- using text analysis to summarize the hundreds of thousands or millions of tweets in each community
- using human expertise to look for likely information themes or tactics in the largest and most central communities.

Figure 2.1 demonstrates this complex methodology at a high level.

Research Approach

Given prior examples of Russian aggression after major sporting events such as the Olympics, we examined whether Russia and its agents might use the 2018 World Cup as an opportunity to conduct malign or subversive information campaigns, potentially as a Phase 0/ Competition Phase preparation for military operations.[2] Our plan was

---

2   We acknowledge that phasing concepts do not map onto Russian strategy and thinking, but use terms and concepts meaningful to our primary U.S. government audience as closest analogs.

Figure 2.1
CLA Method Illustration

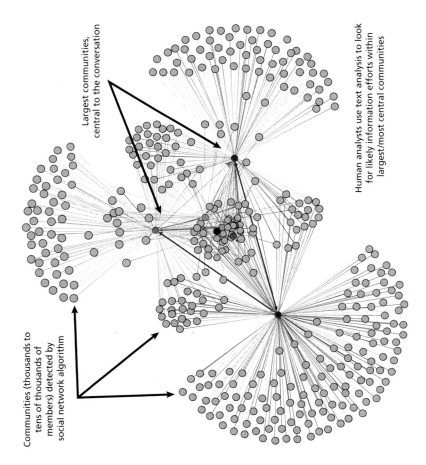

Communities (thousands to
tens of thousands of
members) detected by
social network algorithm

Largest communities,
central to the conversation

Human analysts use text analysis to look
for likely information efforts within
largest/most central communities

to collect the previously described data and conduct CLA. Our goals
were as follows:

- Explore the largest communities in each language using human
  qualitative analysis to contextualize any malign/subversive infor-
  mation-specific findings.
- Look for known/likely Russian malign/subversive information
  campaign themes (e.g., tensions over immigration in France) and
  for known Russian TTPs (e.g., links to Russian state-sponsored
  media like RT).

Experts on our team documented two sets of potential themes to look for based on prior documented Russian malign information efforts. There were two sets of themes:

- broader political themes that included pro-Russia sentiment (e.g., Russia is physically safe to travel in, Russia treats lesbian, gay, bisexual, and transgender populations well), anti-U.S./anti-Western sentiment (e.g., the United States keeps captured Russians from Crimea in prisoner-of-war camps, Western Europe has high levels of crime), support for the Russian annexation of Crimea, retaliation threats for any Ukrainian counteraction, and various anti-Macron accusations (e.g., Macron is secretly owned by Saudis or Jews, he is secretly gay, he has secret financial scandals)

- more sport-specific themes based on large-scale sporting events such as the Olympics and the World Cup, including criticism of the International Olympic Committee and the World Anti-Doping Agency (WADA) for a variety of bans on Russian athletes participating in events, portrayal of Russia as a victim of corrupt international bias, allegations that without Russian participation sporting events are invalid, and broad criticism after bans that the Olympics are a kind of crass political display and represent American/Western hegemony;

- As another potential indicator, use bot detection technology to check bot levels in those communities where we found potential campaigns.

Workflow

Here we outline our workflow in detail, to be used as a template for similar efforts. Though we used Twitter data, any social media data source with account mentions could be used. Similarly, while we used proprietary software, the same general approach could use other software solutions for parts of the process.

1. *Data query:*
   a. After a preliminary round of research on the web, we developed a set of World Cup–relevant terms and parameters to search with, and then employed the preview functionality

of Gnip (Twitter's in-house data vending service) to exam-ine the frequency with which these terms were used among a sample of tweets. This gave us a good idea of the volume of data we could expect in the three languages (English, French, and Russian) that we chose to search in among approximately 69 million tweets.

b. Seventy million tweets and metadata make up a relatively large data set that requires a substantial infrastructure capa-bility (both in terms of computing power and memory) to load, extract, and analyze. We tested a more constrained set of search terms (removing *soccer* and *football* and noncontig-uous uses of *world* and *cup*) to see if we could get a smaller data set, but found that the smaller search term set cut out too much relevant data.

2. *Data pull:* Using the original, more expansive search terms we executed the data pull, which returned just under 69 million tweets.

3. *Extracting, transforming, and loading:*[3]

a. Due to the volume of data, we used a cloud-based server instance with 48 processors and 192 gigabytes of memory. We also employed a range of processes to be efficient with memory. Specifically, we

○ developed an ingestion process with a specific data schema instead of letting the system infer one, which reduced the amount of memory necessary for processing

○ split the data into three smaller/more manageable data sets, one for each language

○ exported the network information to a flat file instead of a relational database, further lowering memory requirements.

2. *Community detection:* We used proprietary software to run a Clauset-Newman-Moore community-detection analysis sepa-rately in each of the three Twitter data sets (English, French, and

[3] This section is idiosyncratic to RAND's software, technology infrastructure, and pro-cesses, but may be useful in helping others plan ahead and learn from our experience.

Russian),[4] allowing us to automatically bin tweets within our text analytics software by community for follow-on characterization of the communities. The broad outlines of this step are as follows:

a. The software first built a mentions network to determine all interactions within the data set.

b. The software then applied a community-detection algorithm to infer communities from the relative frequency of those interactions.

c. Finally, the software loaded the tweets from each community into a data subset for text analysis by a language/culture expert.

5. *Community characterization:* The detection step above discovers the communities in the data set but tells us nothing about the discussion or character of the community. We used a mixed-method analysis of each community's texts, combining human qualitative and machine text analysis.

6. *Network visualization:* The last step in the workflow was to use network visualization software to create a visual representation of the network structure.

Data

We used Gnip to pull 69,007,627 tweets about the 2018 World Cup in a time frame of June 1–July 31, 2018. We searched in three languages (English, French, and Russian) for the following terms:[5]

- World Cup
- #worldcup

---

[4] Aaron Clauset, M. E. J. Newman, and Cristopher Moore, "Finding Community Structure in Very Large Networks," *Physical Review E*, Vol. 70, No. 6, December 2004, Article 066111.

[5] In French, *coupe du monde, #coupedumonde, #cm2018, coupedumonde, jouons à cœur ouvert, jouons à cœur ouvert, and jouons a cœur ouvert.* In Russian, *#чм2018, чемпионат мира, чемпионата мира, чемпионата мира, чемпионату мира, чемпионом мира, чм, футбол, футбола, футболом, футболу, фифа, играй с открытым сердцем, and играть с открытым сердцем.*

- worldcup
- world cup 2018
- world cup 2018 final
- football
- #football
- #FIFAworldcup
- soccer
- #playwithopenheart.

Analytic Methods

For our analysis, we used RAND-Lex, a suite of text analysis and machine-learning tools designed to help human analysts make sense of very large collections of language data.[6] RAND-Lex is a human-in-the-loop approach that leverages what computers do best (work quickly, be reliable, and scale to lots of data) and what humans do best (make meaning and bring context to interpretation). RAND-Lex allows researchers to look at very large collections of text data (i.e., in the tens of millions of words) and conduct both descriptive and exploratory statistical tests to analyze and make meaning of those data sets. In addition to text analysis, RAND-Lex includes social network analysis capabilities.

Network Analysis

We used network analysis to detect communities from the metadata in social media. The community-detection algorithm uses the frequency of interactions to infer social relationships. Individuals who interact (i.e., tweet with one another) on a regular basis are considered to be part of some kind of community. For example, a fan of the New England Patriots professional football team likely interacts more frequently with fans who share the same interest; their frequent interactions come

6 Elizabeth Bodine-Baron et al., *Examining ISIS Support and Opposition Networks on Twitter*, Santa Monica, Calif.: RAND Corporation, RR-1328-RC, 2016; Todd C. Helmus et al., *Russian Social Media Influence: Understanding Russian Propaganda in Eastern Europe*, Santa Monica, Calif.: RAND Corporation, RR-2237-OSD, 2018.

from a shared interest and form the basis for inferring membership in a social group.

For this analysis, we first broke our data into three subsets, one for each language (English, French, and Russian). We then used two methods in our network analysis for each language subset:

- *A community-detection algorithm* to find communities. We used RAND-Lex to sort through all the connections—all the retweets and @'s that make Twitter interactive and social—to build a very large mentions network. We then applied a community-detection algorithm adapted to large networks to infer communities.[7] The inferred communities gave us social groupings to explore, allowing much more insight into the different kinds of communities and stakeholders in the discussion.

- *A network visualization* to see relationships. We used Gephi,[8] a popular network visualization software package, to map out communities and better understand social relationships and interactions over social media. Visualization is important because the network structure can help make clear engagement and relationships between communities.

The social network analysis functioned to reduce the total data set of 69 million tweets that met our query terms down to 34,279,986 tweets that involved social interactions and could therefore be inferred into a network. That reduced total tweet number was further broken down by language:

- English: 29,581,509 tweets
- French: 3,883,056 tweets
- Russian: 815,421 tweets.

This points to an important limitation in our study: Twitter is used unevenly around the world, and thus our Russian data is relatively

---

7   Vincent D. Blondel et al., "Fast Unfolding of Communities in Large Networks," *Journal of Statistical Mechanics: Theory and Experiment*, Vol. 2008, No. 10, 2008, Article 10008.

8   See Gephi, homepage, undated.

sparse as compared with the English and French data.[9] We hope to be able to secure data from more representative and relevant Russian-language platforms in follow-on research.

Text Analysis

For our analysis, we employed three specific text analytic methods:

- *Keyness testing* was used to find conspicuously overrepresented words that signal what a conversation is primarily about. Through this method, a collection of text data (e.g., from a community of interest) is compared with a more general baseline text set (e.g., from other communities) to detect statistically meaningful patterns in word frequencies. If community X uses words such as *sports*, *betting*, *wage*, *line*, and *points* at much higher rates than other groups, we may infer that the group is talking about sports gambling. In this analysis each community was compared with the rest of the data set in order to see what was distinctive about that community relative to the whole.

- *Collocate extraction* was used to find word pairs and triplets, again to understand what the conversation is about. This method looks for words that co-occur close to each other in nonrandom ways, and although they sometimes reflect habitual speech (e.g., "you know"), they also reflect proper names, place names, and abstract concepts. *World* and *cup* co-occur because there is an event named the World Cup. Likewise, *immigrant* and *violence* may co-occur because immigrant violence is an abstract concept some groups talk about.

- *In-context viewing* was used for insight. To better understand conversations, we used RAND-Lex's in-context capability on keywords or collocates to view all the examples in context. For example, the keyword *Putin* is used in many different contexts, so seeing counts for the most frequent phrases with *Putin* in them helped us understand themes in a community.

---

[9] VK is the dominant Russian-language social media and networking platform; see VK, homepage, undated.

Our intent in laying out not only the data and methods used in our analysis but also the workflows used and technical details is to make it possible for others to replicate our work in new contexts and on new data sets. We believe that detecting malign information efforts is a critical challenge for democratic governments and social media companies in liberal democracies, and this report is intended to help advance capability in this area.

CHAPTER THREE

Findings: A Case Study in Detecting Russian Malign or Subversive Information Efforts During the 2018 Fédération Internationale de Football Association World Cup

We positioned our analysis around the 2018 FIFA World Cup, an international soccer tournament held every four years. To do so, we analyzed approximately 69 million tweets in English, French, and Russian about the 2018 World Cup in the month before and the month after the Cup.[1] This analysis allowed us to identify what appear to be two distinct Russian information campaigns,[2] one in Russian and one in French:[3]

- Russian-language efforts were characterized with relatively high confidence as likely including state-sponsored actions. These efforts were focused within two communities.[4]

---

[1] Specifically, we focused on tweets in a time frame of June 1–July 31, 2018.

[2] Russian theory and activities around information (*Informatsionnoye protivoborstvo*) does not directly align with the Department of Defense concept of information operations and information as a joint warfighting function. For the purposes of this report, we refer to detecting malign information *efforts* and *campaigns* that appear to be Russian in origin.

[3] While there was political talk in the English-language data, we did not find indications of malign or subversive information campaigns.

[4] One community, which we labeled Kremlin adepts, appears overall to be state sponsored. A second, which we labeled the pro-Russia community, has evidence of state-sponsored information efforts, operating within an organic community that supports Russia over Ukraine.

- French-language efforts were also focused within two communities. One of these we rated with high confidence, and the other with low confidence, as state sponsored.

While we cannot link these information efforts directly to Russia, we think it likely that the information campaigns we detected as deliberate were indeed conducted by Russia.[5] Russia is actively working to undermine and harm Western democracies through such means, and the themes we detected align with Russian interests in this area.

In the rest of this chapter we describe these two information campaigns[6] and the communities they occurred in, and we provide visualizations of the communities relative to others. We also include a section at the end of the chapter characterizing all ten of the largest communities for the English, French, and Russian data sets.

The World Cup Malign or Subversive Information Campaign: Russian-Language Data

We found a tripartite argument structure in our Russian-language data that included one group of soccer fans and two activist groups using the World Cup as a launching point to argue about the nature and validity of the Russian annexation of Crimea. We characterize the two activist groups as (1) pro-Russia activists supporting annexation and criticizing Ukraine, and (2) Ukraine supporters expressing patriotic support and condemnation of Russia's invasion.

Figure 3.1 demonstrates this social media argument structure. In the figure, the seven largest communities are in color, with the three

---

5   For more on Russia's state-directed information efforts, including Russia's "troll farm," see U.S. House of Representatives Permanent Select Committee on Intelligence, "Exposing Russia's Effort to Sow Discord Online: The Internet Research Agency and Advertisements," webpage, undated.

6   This research was determined by RAND's Human Subjects Protection Committee to be exempt from human subjects protection review, but we still had an obligation to follow ethical principles, as well as to follow our use agreement with Twitter prohibiting identification of individuals. Therefore, we do not quotes tweets from the social media data we collected, as doing so might expose persons to risk. We do cite words and short phrases that were widespread in the data, not pointing to specific individuals.

Figure 3.1
Russian World Cup Argument

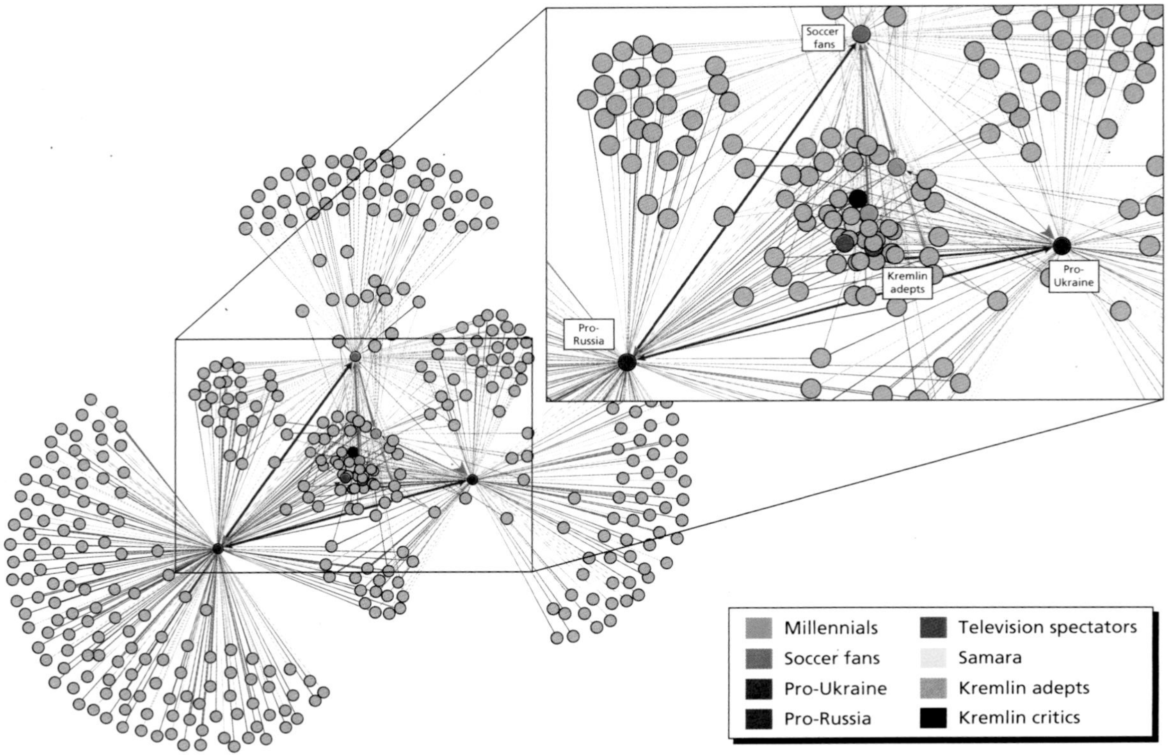

Legend:
- Television spectators
- Samara
- Kremlin adepts
- Kremlin critics
- Millennials
- Soccer fans
- Pro-Ukraine
- Pro-Russia

Inset labels: Pro-Ukraine, Kremlin adepts, Pro-Russia, Soccer fans

groups mentioned above (as well as a Kremlin-oriented community in the center) labeled. We used a social network algorithm to group hundreds of thousands of Twitter accounts into communities based on frequency of user interactions.[7]

- Each dot (node) is a community of Twitter users. The largest ten communities in terms of number of members are shown in color and range in size from thousands to tens of thousands of individual members, while the less-connected communities, shown in gray, are much smaller, ranging from hundreds of members to membership numbers only in the single digits. The most important nodes are the most central, such that they are connected by many lines (edges). Those least important/connected are at the periphery.

- Each edge indicates interactions between communities, and the thicker (higher-weighted) the edge, the more interactions there are. Each edge is an arrow, but most are so small and tenuous that the point of the arrow is invisible. However, for the largest and most central communities, the interactions are so dense that the point of the arrow is visible, showing directionality.[8]

Within this argument structure, we found two communities with apparent Russian information efforts. They are described in detail below.

The Pro-Russia Community

With 298,498 tweets and 26,729 users, the pro-Russia community is the largest in the Russian-language data set, and members are strongly engaged with both the pro-Ukraine and soccer fans communities. In general, this community is characterized by support for Russia and Russian activity, and criticism of Ukraine and the West. The content in this community exceeded our minimum criteria for Russian attribution, including alignment with both specific Russian policy goals and

---

7 We explain the analysis method in more detail in Chapter Two.

8 For example, the edge between the soccer fans community and the pro-Russia community is bidirectional, and relatively thick, with weights of 14,874 and 16,084 in each direction, respectively. By contrast, the pro-Ukraine to soccer fans edge is also bidirectional, but thinner, with weights of 5,395 and 6,625, respectively.

broader interests, as well as multiple Russian TTPs. These include the following:

- Themes and messages supporting Russian policy ends and worldview:

  – One theme is negative political talk around Ukraine and Western countries. Popular tweets in this community are aggressive and inflammatory, including talk about "pro-American terrorists in Syria," critique of WADA's suspension of Russia, contemptuous depictions of Westerners and Western values (e.g., in terms of gender equality). These retweets are part of what is commonly called trolling—that is, disruptive talk meant to antagonize a group.

  – A second theme involves references to President Vladimir Putin threatening Ukraine, noting that if Ukraine initiates military action in Donbas during the World Cup "it would have very serious consequences for Ukrainian statehood."[9] Tweets reporting this statement were positively commented on by other users.

  – A third theme involves the use of #крымнаш (#krymnash), meaning "Crimea is ours," to express support for the annexation of Crimea. The most popular tweet using that hashtag aggressively criticizes a person who condemned the annexation in some of her Facebook posts.

  – Another theme involves a tweet, highly retweeted in this community, that calls Ukrainians fascists, a specific allegation common in Russian propaganda against Ukraine.[10]

- Astroturfing Russian state-supported media activity:

  – The highest key term in this community is @rt_russian, which is the official Twitter account of RT, a well-known channel of Russian propaganda.[11]

---

[9] Denis Pinchuk, Tom Balmforth, and Gabrielle Tétrault-Farber, "Putin Warns Ukraine Against Military Action in East During World Cup," Reuters, June 7, 2018.

[10] Simon Shuster, "Russians Rewrite History to Slur Ukraine over War," *Time*, October 29, 2014.

[11] Helmus et al., 2018, p. 11.

- The most popular RT tweet in this community refers to an article about a Russian singer who was banned from performing at the fans festival in Rostov-on-Don because of her support for the pro-Russian separatists in Eastern Ukraine.[12] Other popular RT tweets in this community contrast the allegedly good situation of Ukrainian prisoners in Russia with the allegedly terrible situation of Russian prisoners in the United States in order to criticize the latter.[13]

- Fake accounts:
  — We found little evidence of fake accounts (either fully automated bots or human-operated troll accounts), and when we used Botometer to score this community for automated accounts, we found a likely 5 percent of accounts that were fully automated.[14]
  — Based on prior research, that is a number we think is within normal bounds.[15] By contrast, the Kremlin adepts community scored for 40 percent of the accounts as bots.
  — We think that a community with primarily authentic accounts and single-digit automated accounts is qualitatively different than a community in which almost half of the voices are automated; the latter seems more like an echo chamber.

Kremlin Adepts

With 20,747 tweets and 5,700 users, the Kremlin adepts is a very centralized community, so named because it focuses official Russian state reporting about the World Cup and events accompanying the tournament. Figure 3.1 shows thick lines between the large pro-Russia community, the pro-Ukraine community, and soccer fans, reflecting social

---

12 James Ellingworth, "FIFA Excludes Russian Singer Who Backed Ukraine Rebels," Associated Press, June 7, 2018.

13 "Russian Pilot Yaroshenko Jailed in US: Conditions in New Prison 'Much Worse,'" Sputnik, June 19, 2018.

14 Botometer is an algorithmic service from Indiana State University that scores Twitter accounts for how likely they are to be automated. In the four communities for which we found likely evidence of malign information efforts we also checked every account in that community through Botometer, reporting the percentages that were found more likely than not to be automated. For more on Botometer, see Botometer, homepage, undated.

15 See Helmus et al., 2018.

engagement between these communities as users reply back and forth. However, the Kremlin adepts community, situated within that triangle of communities, lacks that kind of engagement. So although we found some evidence of Russian influence in the Kremlin adepts community, it appears there was little uptake of its messaging in other communities. This community clearly spreads pro-Russian themes and messages, and we found evidence of three Russian TTPs.[16] The following evidence of Russian influence was discovered:

- Themes and messages supporting Russian policy ends and worldview:
  - Some of the messaging in this community reflects broad pro-Russian/anti-Western sentiment, as well as negative language in reference to Ukraine.
- Astroturfing Russian state-supported media activity
  - This community retweets the narrative of the official social media channels of the Russian administration, including @kremlinrussia (the official account of the Kremlin), @putinrf (the official account of the Russian president), @mid_rf (the official account of the Ministry of Foreign Affairs), and @pravitelstvo_rf (the official account of the Russian government).
  - The most overrepresented keyword for this community is a Twitter handle of a controversial blogger with strong presence in social media who works for the Russian administration. His handle is mentioned 12,708 times, which means that roughly 60 percent of all tweets of this community are retweets of that particular user's tweets—more than an order of magnitude of what we might expect in a community with almost 6,000 users.
- Fake accounts:
  - We found strong evidence of fake accounts in this community. According to the Botometer algorithm, a full 40 percent of the accounts in this community are automated accounts, which is an extremely high number.

---

[16] We did not find much evidence of fake accounts, and Botometer scored this community as having likely 2.5 percent fully automated accounts, which is within normal bounds based on prior research.

Assessing the Engagement Versus the Effectiveness of These Efforts

CLA does not allow for an assessment of messaging effectiveness,[17] but it does show engagement. By *effectiveness* we mean assessing messaging that results in desired behavioral or attitudinal change, and by *engagement* we mean messaging that elicits a response. There is a distinct growing body of research addressing how we might assess both U.S. efforts to inform and adversaries' efforts to influence various audiences.[18]

In this study we were able to look at engagement. We point out the thick edge lines between soccer fans and the pro-Russia and pro-Ukraine communities, indicating high engagement back and forth among the three communities, the highest being between the pro-Russia and soccer fans communities. The Kremlin adepts community, though somewhat integrated with other communities (and thus in the center of the figure), lacks the strong engagement of the pro-Russia community. So while we do not know how effective either effort was, we can say that the pro-Russia community was able to get substantial engagement from the targeted soccer fans community.

The World Cup Malign or Subversive Information Campaign: French-Language Data

We conducted a similar analysis using French-language data and found a four-point, kite-shaped structure comprising what appear to be two authentic soccer fan communities and two overtly political communities with evidence of malign information efforts. These communities are shown in Figure 3.2. We labeled these two communities Paris

17 Resonance analysis allows for the measure of messaging uptake over time and across geography in social media. See Helmus et al., 2018; and William M. Marcellino et al., "Measuring the Popular Resonance of Daesh's Propaganda," *Journal of Strategic Security*, Vol. 10, No. 1, 2017, pp. 32–52.

18 For more on formal assessments of informing, influencing, and persuasion efforts, see Christopher Paul and Miriam Matthews, *The Language of Inform, Influence, and Persuade: Assessment Lexicon and Usage Guide for U.S. European Command Efforts*, Santa Monica, Calif.: RAND Corporation, RR-2655-EUCOM, 2018; and Christopher Paul et al., *Assessing and Evaluating Department of Defense Efforts to Inform, Influence, and Persuade: Handbook for Practitioners*, Santa Monica, Calif.: RAND Corporation, RR-809/2-OSD, 2015.

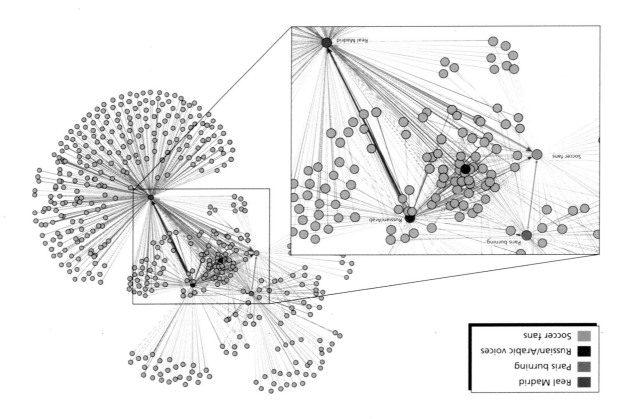

Figure 3.2
French World Cup Argument

burning and Russian/Arabic voices, and both appear to be engaged with the Real Madrid and soccer fans communities. In Figure 3.2, every node comprises a community of between hundreds and tens of thousands of individual members. Our analysis focuses on these four largest communities,[19] which are represented in color in the figure and described in detail below. Edges show interactions between communities, with thickness indicating the density of the interaction.

Within this argument structure, we found two communities with apparent Russian information efforts, described in detail below.

Russian and Arabic Voices

With 701,571 tweets and 192,153 Twitter accounts, the Russian and Arabic voices community is marked by political talk, reflecting both right-wing and left-wing perspectives, with very little soccer content, and while the content was in French, the account profiles were generally in Arabic or Russian. One theme in this community that appears to be authentic was complaints of racism toward people of Arab descent; notably, these complaints came from accounts with Arabic-language profiles. The content in this community exceeded our minimum criteria for Russian attribution, including alignment with both specific Russian policy goals and broader interests, as well as multiple Russian TTPs, including themes that aligned with Russia's informational goal to "exacerbate social tensions in other nations and spread mistrust and confusion." These are as follows:

- Themes and messages supporting Russian policy ends and worldviews (both right-wing and left-wing complaints)
  - populist criticism of French president Macron, including discussion of the *gilets jaunes* movement, which emerged as a rural protest in June 2018 before exploding into violent protests across all of France in November 2018, in reaction to an announced tax raise over fuel

---

- complaints of racism toward immigrants or those of Algerian descent from accounts that appeared to be from Arabic speakers but were tweeting in French[20]

• Accounts that appear to be Russian, tweeting in French
- French-language tweets but Russian-language profiles
- accounts gone within a moth of our analysis, possibly banned by Twitter as bots/trolls

• Fake accounts that appeared to be Russian but were tweeting in French (i.e., trolls/fake accounts), characterized by
- heavy or exclusive retweeting activity (no original content)
- a generic/procedural name with a random number for a handle
- the absence of a profile photo or, conversely a photo too good to be true (e.g., very attractive young women who might be taken for models).[21]

Paris Burning

With 521,348 tweets and 128,216 Twitter accounts, Paris burning is a right-wing community; it is connected to the French conservative party Les Républicains and the identitarian movement on the extreme right of the political spectrum, bearing an anti-immigration stance. The content in this community exceeded our minimum criteria for Russian attribution, including alignment with specific Russian policy goals and broader interests, as well as multiple Russian TTPs, including themes that aligned with Russia's informational goal to "exacerbate social tensions in other nations and spread mistrust and confusion."

This is an example in which ascribing Russian attribution is difficult. On the one hand, the content in this community does align with Russian informational objectives, and we found four distinct themes

---

[20] The Twitter pages themselves are in Arabic, with page headings such as "Following" and "Followers," the user's profile, and other aspects written in Arabic, but the content of the tweets is in French.

[21] Botometer scored this community as having likely 2.5 percent fully automated accounts, a number that is within bounds given previous research.

and an additional TTP (astroturfing Russian media). On the other hand, we did not find evidence of fake accounts,[22] and it is quite possible that the content here reflects authentic French sentiment on the far right. Given that this community does more than exceed our criteria, despite our uncertainty, we detail the TTPs we identified as follows:

- Themes and messages supporting Russian policy ends and worldview:
  - immigrants as a serious problem, responsible for urban violence (and violence against women, in particular)
  - criticism of the Macron administration and allegations about government "silence" over sexual assaults by immigrants after the French victory in the World Cup
  - the notion that FIFA's decision to stop broadcasting close-ups of attractive female supporters during the Cup reflects pandering to Muslims and Qatar (the host of the next World Cup)
  - a theme of *pillaging* [looting] by immigrants, in contrast to *Français de souche* (meaning the nonimmigrant, native French)
- Astroturfing Russian state-supported media activity:
  - the hashtags @sputnik_fr and #sputnikvidéo to mark Sputnik France video content.

Next Steps

Based on our findings and lessons learned in our analysis process, we lay out in Chapter Four a short review of our major points, a road map for operationalizing this method, and a set of specific recommendations for implementation. Given the importance of detecting malign information efforts conducted over social media, we hope the U.S. government can efficiently and quickly implement this or a similar method.

---

[22] We did not find much evidence of fake accounts in this community (e.g., empty profiles with procedural names, missing profile photos, or profile photos of models), and Botometer scored this community as having likely 3.7 percent fully automated accounts, a number well within bounds given previous research.

CHAPTER FOUR

Recommendations and the Way Ahead

By combining scalable network and text analysis software, we were able within days of receipt of data to map out, visualize, and make sense of communities on social media—that is, social groups engaged in argument and advocacy. This method functioned as a kind of data reduction, segmenting data into organic, socioculturally meaningful subsets to rapidly and efficiently search for patterns indicating malign information efforts in data sets that are too large for human qualitative analysis. In particular, our effort addresses a serious U.S. information operations capability gap: scalable detection of malign efforts at the level of aggregate wholes.

In our specific study, within a multilanguage (English, French, and Russian) data set of approximately 69 million tweets, we identified what appear to be Russian information efforts aimed at two distinct audiences. This detection effort was timely—for example, providing early warning of an information effort aimed at right-wing extremists and specifically targeting the anti-Macron, populist *gilets jaunes* movement months before that movement's violent protest in November 2018 gained English-language news attention.

Given the urgency of the need to detect malign and subversive information efforts and given the promise of this proof-of-concept study, we strongly recommend that the United States, allies, and partners consider operationalizing this and complementary methods and technology, as well as the requisite technical and professional expertise. Malign or subversive campaigns are time-sensitive forms of competition. They may present a kind of fait accompli if detected too late. For example, an attempt to influence an election needs to be detected in time for an effective response. Thus there are compelling reasons why the United States and its allies need the capacity to detect these efforts, and a robust ability to detect malign informational efforts is a prerequisite to possible responses. Given this, we offer the following observations and recommendations:

- In our report we have pointed out a serious technical gap in current research on detecting information efforts: current research is focused on detecting individual elements, not wholes at the level of aggregates. We have also offered a model explaining how detection of malign efforts as wholes may be more sensitive to faint signals and more effective than trying to identify efforts at the level of individual accounts or messages. *We recommend that any U.S. government–led efforts to develop information efforts, detection methods, and technology address this research gap.*

- For the foreseeable future, it is unlikely that a purely machine-based approach will have the kind of human-like precision with language data needed to robustly detect malign information efforts. We have outlined a human-in-the-loop approach that leverages what computers do best (work quickly, be reliable, and scale to lots of data) and what humans do best (make meaning and bring context to interpretation of data), maximizing the return of current technology. This approach, however, requires expertise for data ingestion, and the overall process is expert driven. *We recommend that the U.S. government also invest in developing professional expertise to appropriately support early detection of malign information efforts. Such a professional expertise would include data science, social network analysis, text analytics, and cultural/language expertise for select target audiences.*

An effective, agile detection system will require some level of further development. The effort detailed in this report can be seen as prototyping that could with a modest level of effort become a user-friendly end-to-end system that is an important step in countering increasingly aggressive malign information efforts from Russia and other competitors. Given that kind of detection capability, the United States will have the option to respond effectively to this serious threat.

Full Community Characterization

In this appendix we detail the ten largest communities (by volume of tweets) in each language, first noting politically oriented communities, then those that seem genuinely oriented toward sports or other interests, and then within each section by community size in terms of Twitter volume. We point out that the Russian-language Twitter conversation around the World Cup is a fairly political one. While we found only two communities with strong evidence of active information efforts, Twitter is clearly a fertile location for such efforts.

Politically Oriented Russian Communities

Figure A.1 shows the network structure of the Russian-language data set we analyzed. In this report we have focused on the two pro-Russia community accounts with markers of Russian information efforts. For additional context, we offer the community analysis of the remaining top ten communities (according to size).

The Pro-Russia Community (298,498 Tweets/23,974 Users)

While the pro-Russia community appears to predominantly praise the Russian soccer national team and Russian government for the organization of the 2018 World Cup, there are multiple popular (often retweeted) tweets written in a negative political tone regarding Ukraine and Western countries. For example, a popular message in this community is a statement made by President Vladimir Putin threatening Ukraine, stating that if Ukraine initiated military action in Donbas

Figure A.1
Russian World Cup Argument

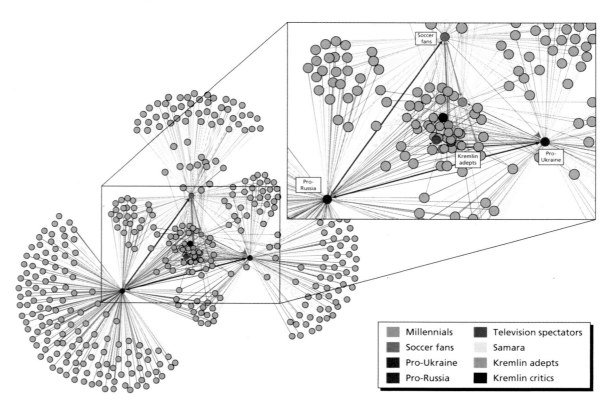

Millennials   Television spectators
Soccer fans   Samara
Pro-Ukraine   Kremlin adepts
Pro-Russia   Kremlin critics

during the World Cup "it would have very serious consequences for Ukrainian statehood."[1] This community uses #крымнаш (#krymnash, meaning "Crimea is ours") with frequency, which is a strong indicator of support for the Kremlin's policy toward the Crimean Peninsula and Ukraine. Another popular tweet calls Ukrainians fascists, which is common in Russian propaganda.[2]

Keyness testing revealed that the most overrepresented keywords in this community are Twitter handles of users involved in spreading pro-Kremlin narrative, including @rt_russian (at first place on the list), which is the official Twitter account of RT in Russia. The most popular RT tweet in this community refers to an article about a Russian singer who was banned from performing at a fan festival in Rostov-on-Don because of her support for the pro-Russian separatists in Eastern Ukraine.[3] Other popular tweets in this community contrast the allegedly good situation of Ukrainian prisoners in Russia with the allegedly terrible situation of Russian prisoners in the United States in order to criticize the latter.[4] Other popular tweets involving political trolling include talk about pro-American terrorists in Syria, critique of WADA, and contemptuous depictions of Westerners and Western values (e.g., in terms of gender equality).

The Pro-Ukraine Community (186,166 Tweets/26,729 Users)

Keyness testing revealed underrepresentation of keywords and hashtags related to the World Cup, such as #чм2018 (#wc2018), #worldcup, @sportsru, @fifaworldcup_ru, *Russian team*, *match*, and *goal*. Even when these keywords and hashtags are used by the pro-Ukraine community, they serve only as a pretext to criticize the Russian government and support Ukraine. One of the most popular tweets in this community prizes a young Ukrainian boy who reportedly was killed in Eastern Ukraine for wearing the Ukrainian flag. There appears to be an over-

1   Pinchuk, Balmforth, and Tétrault-Farber, 2018.

2   Shuster, 2014.

3   Ellingworth, 2018.

4   "Russian Pilot Yaroshenko Jailed in US," 2018.

representation of such keywords and hashtags as *Ukraine*, *Pussy Riot* (the antigovernment Russian activist punk band), *slava Ukraine* (glory to Ukraine), *war*, *military*, and #mh17.

Keyness testing revealed that certain Twitter handles are among the most overrepresented keywords and hashtags in this community, including @fake_midrf (at first place on the list), a parody of the Russian Ministry of Foreign Affairs account. In addition, users' most popular tweets in this community speculate about the death of Oleh Sentsov, a Ukrainian film director detained in a Russian prison who went on a hunger strike. Other popular tweets prize the band Pussy Riot for disrupting the World Cup final by running onto the pitch and a Croatian player for exclaiming "Glory to Ukraine" after his team won over Russia in the World Cup quarterfinals.

Kremlin Adepts (20,747 Tweets/5,702 Users)

The Kremlin adepts community retweets narratives of official social media channels of the Russian administration. Overrepresented keywords and hashtags we found include @kremlinrussia (the official account of the Kremlin), @putinrf (the official account of the Russian president), @mid_rf (the official account of the Ministry of Foreign Affairs), and @pravitelstvo_rf (the official account of the Russian government). The most overrepresented keyword for this community is the Twitter handle of a controversial blogger with a strong presence in social media who works for the Russian administration; his handle is mentioned 12,708 times in this community, which means that roughly 60 percent of all community tweets are retweets of that particular user. It is a very centralized community, concentrated around official Russian narratives about the World Cup and events accompanying the tournament.

Domestic Situation Critics (15,573 Tweets/5,828 Users)

Based on keyness testing coupled with collocation analysis, the domestic situation critics community predominately shares YouTube videos—@youtube is the most overrepresented keyword in this community—that criticize Russian authorities for adopting new regulations during the World Cup, the Russian people for the ways they take financial advantage of tourists and soccer fans, or the general domestic situation. Some members of this group seem to be Ukrainians, since they are using social media in Ukrainian (the collocation мені подобається відео, meaning "I liked the video" in Ukrainian, appeared over 600 times). At the same time, this community is not particularly concerned about the situation in Ukraine. While there are some tweets supportive of Ukraine, the strongest focus lies on the Russian domestic situation.

Communists (13,242 Tweets/2,674 Users)

Retweeting users in the communists community tended to either describe themselves as communist or post communist content. This community appears to have negative views on the current Russian regime for denying its communist past (e.g., outrage about the lack of the red star from the Kremlin's Spasskaya Tower on the official FIFA 2018 screen saver for the tournament), for privileges of elites and their lack of patriotism, and for increasing costs of living. Nevertheless, they do not share any pro-Ukrainian messages. They are strongly focused on domestic problems and hope for a communist solution to Russia's social and economic issues.

Russian Sports- and Other-Oriented Communities

Soccer Fans (158,734 Tweets/27,830 Users)

Text analysis did not reveal any distinct political stance of the soccer fans group. Keywords associated with stances represented by the two previous groups are evenly underrepresented in this community. Members of this community sometimes retweet pro-Ukrainian or anti-Western messages, but very rarely. There is no significant praise or critique of the Russian government. Instead there is an overrepresentation of hashtags associated directly with soccer, such as #worldcup and #чм2018 (#wc2018), as well as the handles of sports channels, websites, and commentators and bloggers, such as @sportsru, @eurosport_ru, @sportexpress, and @championat. One example of a popular tweet in this community is, "Forget the Final. The Most Memorable Moment of the World Cup 2018—Fernandez Goal in Match with Croats."[5] It seems that this is a

---

5   Sports.ru (@sports.ru), "Забудьте о финале. Главный момент ЧМ-2018—гол Фернандеса хорватам [Forget the Final. The Most Memorable Moment of the World Cup 2018—Fernandez Goal in Match with Croats]," Twitter post, July 16, 2018.

community of users who are interested in sports in general and in soccer in particular.

Millennials (62,561 Tweets/31,001 Users)

The millennial community talks about soccer rather ironically, but without any political stance. It seems that it is a community of young people who are not particularly interested in sports but follow the World Cup as a contemporary event. They do not present any political views. Keyness testing revealed that the most overrepresented keywords in this community are Twitter handles of college students and young artists, who refer to the World Cup with sarcastic comments. The tournament is rather a pretext to share thoughts on loosely related topics.

Television Spectators (35,688 Tweets/5,840 Users)

None of the overrepresented keywords in the television spectators community that were revealed by keyness testing carry any political meaning. The community mainly discusses the championship, but contrary to the soccer fans community, it seems that this discussion is motivated by the event of the World Cup rather than a genuine interest in soccer. For example, overrepresented keywords include the Twitter handles of @fifaworldcup_ru (at first place on the list) and @uefacom_ru, which are official accounts in Russian for the World Cup and the Union of European Football Associations. Other overrepresented handles are @channelone_rus (one of the main Russian television stations), @urgant_show (a late-night television show), @volgogradtrv (a local branch of a big television network), @fcsm_official (the FC Spartak Moscow football team), and @otkritiearena (the stadium of FC Spartak Moscow).

Residents of Samara (23,172 Tweets/2,665 Users)

Keyness testing unambiguously demonstrates that the distinct feature of the residents of Samara is defined geographically. The most overrepresented keyword for this community is *Samara*, which was one of the cities hosting the World Cup. The second most overrepresented keyword is the Twitter handle of the governor of Samara (@D_Azaroff). Popular tweets in this community share local experiences of the World Cup: they prize opening a fan zone and temporary festival on one of the main streets of Samara and share the stories about soccer fans visiting the city.

The Meme Community (10,040 Tweets/4,566 Users)

Users in the meme community tend to post memes, humorous videos, pictures, and stories. The World Cup appears to be another theme of this type of content. The community represents no particular political stance. Those in the community both make fun of and show pride for the Russian national team. One popular tweet mentions the decision of Russian government to raise retirement age, taken during the World Cup, but there is no direct criticism of the government. *Putin* is an underrepresented keyword in this community.

French Political Communities

Figure A.2 shows the network structure of the French-language data set we analyzed. In the main report, we focused on the two accounts with markers of Russian information efforts. For additional context, we offer the community analysis of the remaining top ten communities (according to size).

Russian and Arabic Voices (70,571 Tweets/192,152 Users)

The Russian and Arabic voices community only discusses soccer occasionally. Most of the overrepresented keywords are unrelated to soccer. Members of this community represent a rather heterogeneous group that tweets about various topics from rap music to Japanese graphic novels. They discuss current events, such as the *gilets jaunes* movement blocking roads in several parts of France to oppose President Macron's tax increase on gas. While this does not appear to be an overly political community, *Macron* is an overrepresented keyword that appears in 135 different tweets, of which some are innocuous and some critical of the president.[6]

An interesting feature of this community is the presence among overrepresented keywords of Twitter accounts in Russian, some of which are likely bots, and in Arabic (meaning that the Twitter pages are not in French, although the tweets themselves are written in French). For

---

[6] It is worth noting, however, that as of November 2018, Macron's approval rate was only 25 percent, meaning that a critical view of Macron is probably very common on Twitter and other social media; "La popularité de Macron continue de chuter [The Popularity of Macron Continues to Fall]," *Les Échos*, November 18, 2018.

Figure A.2
French World Cup Argument

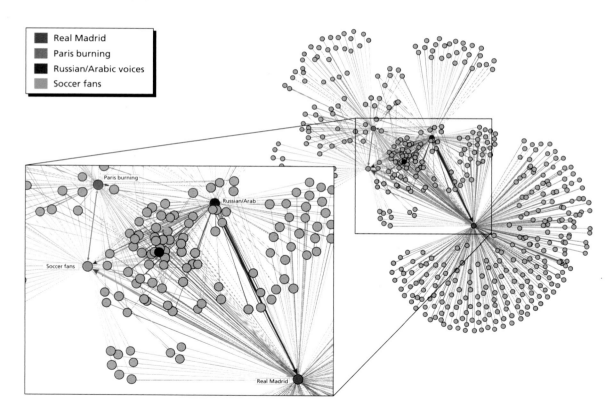

Legend:
- Real Madrid
- Paris burning
- Russian/Arabic voices
- Soccer fans

Russian pages, a tweet from one of these users has a patriotic connotation (although a rather innocuous one; she indicated she would scream with her French flag during the World Cup); another reported a racist aggression during the celebrations of the French World Cup against people waving an Algerian flag; another tweet seems to refer to immigrants of Spanish or Portuguese descent who can publicize their origin while others (perhaps in reference to the Algerian flag incident above) cannot. This last tweet was retweeted in French-language accounts on Russian pages, some of which appear to be trolls based on their exclusive retweeting activity. One of these French-speaking Russian accounts indicates a proclivity toward La France Insoumise (the extreme left-wing political party of Jean-Luc Mélenchon, who came in fourth in the 2017 presidential elections). We could not identify a particular common thread among overrepresented Twitter accounts in Arabic, except for discussions of the *gilets jaunes* movement.

Paris Burning (521,348 Tweets/128,216 Users)

Based on our lexical analysis, the Paris burning community is focused on sharing political views, not soccer. A recurrent theme in this community appears to be addressing urban violence (and violence against women, in particular), with immigrants being identified as the culprits in a number of cases. This community appears to lean toward the right—and sometimes extreme right—side of the political spectrum.

One overrepresented user among this community is a parliamentarian from the French conservative party Les Républicains who is very critical of French president Macron and his junior minister, Secretary of State Marlène Schiappa, on gender parity and nondiscrimination. The most shared tweets from this user mention the "silence" of the government on the sexual aggressions that took place during the celebrations after France won the World Cup (several tweets, 1,740 counts for the most shared). Another much-shared tweet (at 1,932 counts) is about FIFA's decision to stop filming close-ups of attractive female supporters during games, with this decision being presented as censoring women based on how they dress or look. Some comments establish a link between this decision and the fact that the next World Cup is planned to take place in Qatar. Other overrepresented keywords include *pillage* and *pillages* [looting] and refer to street violence incidents that took place in Lyon and Paris, in particular, following France's World Cup victory. *Pillage* is mentioned in different tweets, shared many times, often citing mainstream media reports (e.g., @bfmtv and @europe1). The tweet on *pillage* with the highest count (at 1,058) was, however, authored by a user (@lugus69) who appears to be part of the identitarian movement, which is on the extreme right of the political spectrum and has an anti-immigration stance.

Other interesting characteristics of this community include the presence of the French president among the most overrepresented keywords, with 1,255 different mentions, usually with very low counts. Another most overrepresented keyword is @sputnik_fr, which is mentioned in 393 different tweets, usually retweets from the Sputnik website, and mostly, but not exclusively, on the World Cup. For instance, one retweet mentions Russian diplomats; another mentions British elected officials.[7] #*Poutine* is also among the overrepresented keywords, mainly for retweets of a single tweet from U.S. president Donald Trump congratulating the Russian president for his "very beautiful World Cup, one of the best in History." Finally, among the most overrepresented keywords are @frdesouche and @f_desouche, which refer to *Français de souche* (meaning nonimmigrant, native French). These Twitter handles, which are very strongly positioned at the extreme right of the French political spectrum, are mentioned on various occasions with significant counts (often through retweets), suggesting that they are routine participants in this community.

French Sports- and Other-Oriented Communities

Real Madrid (1,438,822 Tweets/105,021 Users)

The Real Madrid community is focused almost exclusively on soccer, and more specifically the Real Madrid soccer club. Twitter accounts devoted to soccer-related news and with large numbers of followers (@footballogue and @actufoot) are overrepresented among the mentions and retweets of this community. Also overrepresented are terms

7   However, the full content of retweets from @sputnik_fr could not be retrieved from Google searches, making the identification of more political tweets difficult.

and hashtags related to soccer, such as #cm2018 (with "cm" standing for Coupe du Monde, the French name for the World Cup), names of players, abbreviations for countries (e.g., "BR" to indicate a game involving Brazil), or keywords like *officiel* [official], *supporter*, or *qual-ifiés* [qualifies]. Furthermore, this community seems to be predominantly supporting, or interested in, Real Madrid. Overrepresented keywords include names of players who play or might transfer to Real Madrid, and also prominent are the French official Twitter account of Real Madrid and the Twitter account of a user (with 26,200 followers) who tweets mostly about soccer but seems to have a preference for Real Madrid, with occasional retweets of a more political nature.

Soccer Fans (476,174 Tweets/68,360 Users)

The soccer fans community is almost exclusively focused on soccer news and commentary. The most overrepresented keyword in this community is a Twitter account that provides statistics about soccer: @optajean. This community's favorite media are some of the most popular sports news outlets, including Radio Monte Carlo, which is a popular French radio network with a heavy sports news content; Eurosport, a television channel exclusively focused on sports; and *L'Équipe*, which is the most popular daily sports newspaper; all three appear in the top five most overrepresented keywords for this community. Other overrepresented keywords include various Eurosport and Radio Monte Carlo shows (@afterrmc and #rmclive) and journalists. This community appears to be particularly interested in the AS Monaco soccer club, which is the only overrepresented club name, and was tweeted or retweeted 108 times.

The International Soccer Community (220,610 Tweets/35,099 Users)

All overrepresented keywords in the international soccer community relate to soccer, suggesting that this is the only topic that this community really discusses. We identified five distinct threads, which are difficult to reconcile:

1. A number of overrepresented keywords and hashtags refer to the soccer club of the Barcelona FC soccer club (@fc_barcelona, @fc_barcelona_fra, #barçaworldcup, and barça) and Twitter accounts of players from FC Barcelona.

2. Other overrepresented keywords and hashtags are the Twitter account for the Belgian national team, the Red Devils (@belreddevils, #diablesrtbf, #redtogether, and *diables* [devils]); Belgian soccer players; Belgian media (@RTBFsport and @cfoortrbf); and Belgium as a country (*belgique*), suggesting that at least part of this community is mostly Belgian or located in francophone Belgium.

3. A third thread in this community is around the main newspaper read in southern France, *Midi Libre*, with mentions of its Twitter accounts (@midilibresports and @midilibre) and its journalists.

4. There is also a Senegalese thread, with several Senegalese or Senegal-based users' names appearing among the most overrepresented keywords. One of these users is a Senegalese engineer with 12,400 followers who discusses political issues, with a specific focus on energy policy. A view in context analysis shows that this user is frequently retweeted by this community. Other overrepresented keywords include the Twitter account of @AfricaFootDaily, which provides news on African soccer teams and games, as well as @senegalais and #senegal.

5. Finally, some overrepresented keywords also refer to Brazil in the World Cup or Brazilian players (*brésiliens* and @cbf_futebol).

While it is difficult to identify a common thread between these five themes, the discussions that take place among this community appear to focus on soccer, with relatively few political references.

Soccer Fans/Spanish Clubs (144,405 Tweets/34,731 Users)

The soccer fans/Spanish clubs community is focused on the World Cup, with no evidence of specific political proclivities or discussions with a political content. Based on our lexical analysis, what sets it apart from others is its heavy reliance on various media for news and information about the World Cup, which are at the same time mainstream (in the sense of being some of the most popular media for this purpose) and different from the ones in the community of soccer fans. These overrepresented media are TF1 (@tf1 is the most overrepresented keyword for this community), which is the French television channel with the highest number of viewers;[8] RTL, which is the number one French radio network in terms of numbers of listeners (@rtlfrance);[9] and beIN Sports (@beinsports_fr), which is a sports channel belonging to the Al Jazeera group. Other overrepresented keywords include the names of a former French soccer player who covered the World Cup for both RTL and TF1, journalists from the media outlets mentioned above, and others (not all of them covering the World Cup). A number of soccer players' names are also among overrepresented keywords; they appear to be either players who were part of the French team during the World Cup, players in Spanish clubs (Atlético and Real Madrid), or both.

The Sports Betting Community (133,126 Tweets / 13,122 Users)

The sports betting community appears to be predominately composed of soccer fans who have an interest in betting on the results of games. The most overrepresented keyword in this community is the Twitter handle of a mobile phone application for sports bets (@unibetfrance), and the third most overrepresented hashtag is #freebet (we also find the keywords *freebet* and *freebets* further down the list). Other Twitter handles and hashtags related to betting (such as the online betting site @betclic) appear among the most overrepresented keywords, as do more common words such as *tirage* [drawing, as in a drawing for a lottery], *gagner* [win], and *jeu* [game, but not in the sense of a soccer match]. Other types of overrepresented keywords and hashtags are shortcuts for country names (e.g., #uru for Uruguay and #por for Portugal) or shortcuts for games (e.g., #cronig for Croatia versus Nigeria or #uruksa for Uruguay versus Saudi Arabia) that are used by members of the community to discuss, and bet on, World Cup games. This community appears primarily focused on sports, with little to no political discussions.

---

8   This was as of January 2018. See, for instance, "Audiences 2017: TF1 en recul à 20%, mais domine toujours France 2 et M6 [TF1 Down to 20%, but Still Dominates France 2 and M6]," Europe 1, January 2, 2018.

9   This was as of January 2018. See, for instance, "Audiences radio: RTL reste la radio la plus écoutée, Europe 1 au plus bas [RTL Still the Most Listened to Radio, Europe 1 the Lowest]," Sud Ouest, January 18, 2018.

The Contestant Community (126,681 Tweets/22,800 Users)

The contestant community seems to focus on games and contests for which prizes can be won. The most overrepresented keywords include *gagner* [win], *concours* [contest], *jeu* [game], #jeuconcours, *chance* [luck], *remporter* [win], #concours, *cadeau* [gift], *collector*, and *participer* [participate]. Other overrepresented keywords include the names of Twitter accounts devoted to video gaming and mobile gaming apps (@ggstories and @prizee_officiel), as well as products and brands that appear to be prizes to be won in such contests (@polkaudio, *asus*, *smartphone*, *black-berry*, @bluemicrophones, *casques* [headsets], @speckproducts, *coques* [smartphone cases], #samsung, and #galaxys9). This community does not appear to discuss the World Cup or soccer in general much.

The Olympique de Marseille Community (88,978 Tweets/13,640 Users)

Based on our lexical analysis, it appears that this community is mostly composed of fans of the Olympique de Marseille (OM) soccer club from Marseille. Three out of the five most overrepresented keywords are the twitter handles of self-declared OM fans, who tweet mostly about the club's players and games. The most overrepresented keyword is the official Twitter account of the club, and the fourth is the Twitter handle of an OM player (also on the French national team). The name of this player comes up two more times in the most overrepresented tweets. Another most overrepresented keyword is the hashtag #teamom (team OM). Discussions within this community seem to focus on soccer, and specifically the OM, with little to no political content.

The YouTube and Gaming Community (31,341 Tweets/12,517 Users)

Based on our lexical analysis, the YouTube and gaming community appears to be particularly active on YouTube, either by creating content or sharing videos. Its members also seem to share an interest in video games. The most overrepresented keyword in this community is @youtube, which is mentioned in almost 6,000 tweets with various counts. Most often it comes up within a sentence that starts with "j'aime une video @youtube" [I like a @youtube video] followed by the link to that video. In other cases, it is mentioned in the context of a YouTube playlist, or in "via @youtube" to indicate the provenance of a video mentioned in a tweet. Based on the first words of the titles of the videos, as they appear in a "view in context" search for @YouTube, the videos mentioned in this community's tweets are overwhelmingly World Cup and soccer related, with no indications of political content. Oddly, one of the most overrepresented keywords is @efs_dondesang, which is the French authority that collects blood donations. This comes as the result of two tweets that are much retweeted: one from a user who is very active posting videos on YouTube (and has 1.39 million subscribers), and the other (from an association working on soccer and disabilities retweeting a call from @efs_dondesang noting that the number of people coming to their centers to donate blood had decreased because of the heat and the World Cup and calling on donors to show up. We also find a number of Twitter handles related to video games among the most overrepresented keywords, such as @instangaming.fr, #instantgamingworldcup, or @nintendofrance; and users very active on YouTube, writing on video games, or both.

References

Al-Khateeb, Samer, and Nitin Agarwal, "Understanding Strategic Information Manoeuvres in Network Media to Advance Cyber Operations: A Case Study Analysing Pro-Russian Separatists' Cyber Information Operations in Crimean Water Crisis," *Journal on Baltic Security*, Vol. 2, No. 1, 2016, pp. 6–27.

"Audiences 2017: TF1 en recul à 20%, mais domine toujours France 2 et M6 [TF1 Down to 20%, but Still Dominates France 2 and M6]," Europe 1, January 2, 2018. As of September 17, 2019:
http://www.europe1.fr/medias-tele/audiences-2017-tf1-en-recul-a-20-mais-domine
-toujours-france-2-et-m6-3535340

"Audiences radio: RTL reste la radio la plus écoutée, Europe 1 au plus bas [RTL Still the Most Listened to Radio, Europe 1 the Lowest]," Sud Ouest, January 18, 2018. As of September 17, 2019:
https://www.sudouest.fr/2018/01/18/audiences-radio-rtl-reste-la-radio-la-plus-ecoutee
-europe-1-au-plus-bas-4214491-10228.php

Beehner, Lionel, Liam Collins, Steve Ferenzi, Robert Person, and A. F. Brantly, *Analyzing the Russian Way of War: Evidence from the 2008 Conflict with Georgia*, West Point, N.Y.: Modern War Institute at West Point, March 20, 2018. As of April 9, 2019:
https://mwi.usma.edu/wp-content/uploads/2018/03/Analyzing-the-Russian-Way
-of-War.pdf

Blondel, Vincent D., Jean-Loup Guillaume, Renaud Lambiotte, and Etienne Lefebvre, "Fast Unfolding of Communities in Large Networks," *Journal of Statistical Mechanics: Theory and Experiment*, Vol. 2008, No. 10, 2008, Article 10008.

Bodine-Baron, Elizabeth, Todd C. Helmus, Madeline Magnuson, and Zev Winkelman, *Examining ISIS Support and Opposition Networks on Twitter*, Santa Monica, Calif.: RAND Corporation, RR-1328-RC, 2016. As of December 8, 2018:
https://www.rand.org/pubs/research_reports/RR1328.html

Botometer, homepage, undated. As of September 17, 2019:
https://botometer.iuni.iu.edu/#!/

49

Casebeer, William D., "Narrative Technology to Detect and Defeat Adversary Ideological Influence," in Mariah Yager, ed., *SMA White Paper: What Do Others Think and How Do We Know What They Are Thinking?* Washington, D.C.: U.S. Department of Defense and Joint Chiefs of Staff, March 2018, pp. 129–138.

Chew, P. A., and J. G. Turnley, "Understanding Russian Information Operations Using Unsupervised Multilingual Topic Modeling," in Dongwon Lee, Yu-Ru Lin, Nathaniel Osgood, and Robert Thomson, eds., *Social, Cultural, and Behavioral Modeling: 10th International Conference, SBP-BRiMS 2017, Washington, DC, USA, July 5–8, 2017, Proceedings,* Cham, Switzerland: Springer International Publishing, 2017, pp. 102–107.

Clauset, Aaron, M. E. J. Newman, and Cristopher Moore, "Finding Community Structure in Very Large Networks," *Physical Review E,* Vol. 70, No. 6, December 2004, Article 066111.

DiResta, Renee, Kris Shaffer, Becky Ruppel, David Sullivan, Robert Matney, Ryan Fox, Jonathan Albright, and Ben Johnson, *The Tactics & Tropes of the Internet Research Agency,* New York: New Knowledge Organization, 2018.

Dowse, Suzanne, "Mega Sports Events as Political Tools: A Case Study of South Africa's Hosting of the 2010 FIFA World Cup," in J. Simon Rofe, ed., *Sport and Diplomacy: Games Within Games,* Manchester, England: Manchester University Press, 2018, pp. 70–86.

Ellingworth, James, "FIFA Excludes Russian Singer Who Backed Ukraine Rebels," Associated Press, June 7, 2018. As of November 28, 2018:
https://www.apnews.com/84a0d69aa6c54e798b4b7e58de66c6af

Gephi, homepage, undated. As of September 16, 2019:
https://www.gephi.org

Grix, Jonathan, and Nina Kramareva, "The Sochi Winter Olympics and Russia's Unique Soft Power Strategy," *Sport in Society,* Vol. 20, No. 4, 2017, 461–475.

Helmus, Todd C., Elizabeth Bodine-Baron, Andrew Radin, Madeline Magnuson, Joshua Mendelsohn, William Marcellino, Andriy Bega, and Zev Winkelman, *Russian Social Media Influence: Understanding Russian Propaganda in Eastern Europe,* Santa Monica, Calif.: RAND Corporation, RR-2237-OSD, 2018. As of December 8, 2018:
https://www.rand.org/pubs/research_reports/RR2237.html

"La popularité de Macron continue de chuter [The Popularity of Macron Continues to Fall]," *Les Échos,* November 18, 2018. As of September 17, 2019:
https://www.lesechos.fr/politique-societe/emmanuel-macron-president/0600168024159-la-popularite-de-macron-continue-de-chuter-2222650.php

Marcellino, William M., Kim Cragin, Joshua Mendelsohn, Andrew Micahel Cady, Madeline Magnuson, and Kathleen Reedy, "Measuring the Popular Resonance of Daesh's Propaganda," *Journal of Strategic Security,* Vol. 10, No. 1, 2017, pp. 32–52.

Matthews, Miriam, Alyssa Demus, Elina Treyger, Marek Posard, Hilary Smith, and Christopher Paul, *Russian Social Media Influence: Understanding and Defending Against Russia's Malign and Subversive Information Efforts in Europe*, Santa Monica, Calif.: RAND Corporation, RR-3160-EUCOM, forthcoming.

Paul, Christopher, and Miriam Matthews, *The Language of Inform, Influence, and Persuade: Assessment Lexicon and Usage Guide for U.S. European Command Efforts*, Santa Monica, Calif.: RAND Corporation, RR-2655-EUCOM, 2018. As of March 12, 2019:
https://www.rand.org/pubs/research_reports/RR2655.html

Paul, Christopher, Jessica Yeats, Colin P. Clarke, Miriam Matthews, and Lauren Skrabala, *Assessing and Evaluating Department of Defense Efforts to Inform, Influence, and Persuade: Handbook for Practitioners*, Santa Monica, Calif.: RAND Corporation, RR-809/2-OSD, 2015. As of April 1, 2019:
https://www.rand.org/pubs/research_reports/RR809z2.html

Pinchuk, Denis, Tom Balmforth, and Gabrielle Tétrault-Farber, "Putin Warns Ukraine Against Military Action in East During World Cup," Reuters, June 7, 2018. As of November 20, 2018:
https://www.reuters.com/article/us-russia-putin-ukraine/putin-warns-ukraine-against-military-action-in-east-during-world-cup-idUSKCN1J31G5

Rofe, J. Simon, "Introduction: Establishing the Field of Play," in J. Simon Rofe, ed., *Sport and Diplomacy: Games Within Games*, Manchester, England: Manchester University Press, 2018, pp. 1–10.

"Russian Pilot Yaroshenko Jailed in US: Conditions in New Prison 'Much Worse,'" Sputnik, June 19, 2018. As of January 28, 2019:
https://sputniknews.com/us/201806191065533889-usa-russia-yaroshenko-jail-conditions/

Shuster, Simon, "Russians Rewrite History to Slur Ukraine over War," *Time*, October 29, 2014. As of January 28, 2019:
http://www.time.com/3545855/russia-ukraine-war-history/

Sports.ru (@sports.ru), "Забудьте о финале. Главный момент ЧМ-2018—гол Фернандеса хорватам [Forget the Final. The Most Memorable Moment of the World Cup 2018—Fernandez Goal in Match with Croats]," Twitter post, July 16, 2018. As of September 17, 2019:
https://twitter.com/sportsru/status/1018902808270163968

Svärd, Mikael, and Philip Rumman, "Combating Disinformation: Detecting Fake News with Linguistic Models and Classification Algorithms," thesis, Stockholm: KTH Royal Institute of Technology, 2017.

Thomas, Timothy L., "Dialectical Versus Empirical Thinking: Ten Key Elements of the Russian Understanding of Information Operations," *Journal of Slavic Military Studies*, Vol. 11, No. 1, 1998, pp. 40–62.

Träng, David, Fredrik Johansson, and Magnus Rosell, "Evaluating Algorithms for Detection of Compromised Social Media User Accounts," in *ENIC '15: Proceedings of the 2015 Second European Network Intelligence Conference*, Washington, D.C.: IEEE Computer Society, 2015, pp. 75–82.

Tulli, Umberto, "'They Used Americana, All Painted and Polished, to Make the Enormous Impression They Did': Selling the Reagan Revolution Through the 1984 Olympic Games," in J. Simon Rofe, ed., *Sport and Diplomacy: Games Within Games*, Manchester, England: Manchester University Press, 2018, pp. 223–242.

U.S. House of Representatives Permanent Select Committee on Intelligence, "Exposing Russia's Effort to Sow Discord Online: The Internet Research Agency and Advertisements," webpage, undated. As of September 17, 2019:
https://intelligence.house.gov/social-media-content/

VK, homepage, undated. As of September 17, 2019:
http://www.vk.com